SCIENCE MUSEUM

PLASTICS

John Brydson

London: HMSO

ISBN 0 11 290478 5

First published 1991

The photographs in this book are © Science Museum, with the exception of: Amstrad plc (word processor, front cover and p 21); BXL Plastics Ltd/John Acres (extrusion line, p 23); Electrolux (vacuum cleaner, p 32); ERF Trucks (lorry cabin, p 12); ICI plc (squash court, p 35); InterCity (diesel locomotive, p 26); Kenwood Ltd (food processor, p 17); Philips Electronics (kettle jug, p 9); Rubber and Plastics Processing Industry Training Board (figure 2, p 22 – from *Injection Moulding*, 1978, p 10); Salomon Ski Equipment (Great Britain) Ltd (ski boot, front cover and p 5). The author is grateful to the above for permission to reproduce their photographs.

HMSO publications are available from:

HMSO Publications Centre
(Mail and telephone orders only)
PO Box 276, London, SW8 5DT
Telephone orders 071-873 9090
General enquiries 071-873 00111
(queuing system in operation for both numbers)

HMSO Bookshops
49 High Holborn, London, WC1V 6HB 071-873 0011 (counter service only)
258 Broad Street, Birmingham, B1 2HE 021-643 3740
Southey House, 33 Wine Street, Bristol, BS1 2BQ 0272 264306
9–21 Princess Street, Manchester, M60 8AS 061-834 7201
80 Chichester Street, Belfast, BT1 4JY 0232 238451
71 Lothian Road, Edinburgh, EH3 9AZ 031-228 4181

HMSO's Accredited Agents
(see Yellow Pages)

and through good booksellers

Contents

1 Plastics Today

It is difficult to think of the world as we know it today without plastics. If we had been born two hundred years ago, no matter how wealthy, our knowledge of the world would have been very limited. Communications were very slow for there were no telephone, radio or television. There were no photographs, and knowledge of what people and places looked like was limited either to personal contact or to paintings and drawings. There were no gramophone records or magnetic tapes, and outside the big cities there were few opportunities to hear a wide range of music. All of these objects which we take for granted today have one thing in common – they are now made in whole or in part from plastics.

Without doubt plastics have also made travel much easier. Cars produced early this century had few plastics parts, but today would be prohibitively costly to buy and to run and also lack the performance and comfort of modern vehicles. In a modern car we find a variety of plastics: under the bonnet in the form of carburettor floats, cooling fans, distributor caps; in the interior for upholstery, fascia panels, door armrests and roof linings; and externally for the radiator grilles and front- and rear-end bumpers. As I type these sentences on a word-processor, itself largely made of plastics, I am looking at a photograph of a small car. The caption underneath the photograph tells me that out of the 2,730 parts from which the car is assembled, 771 are of plastics. The use of plastics is of course not confined to land-based means of transport. Glass-reinforced plastics are used not only for pleasure craft but also for warships such as minesweepers. Owing to their high strength–weight ratio, reinforced plastics are also widely used in aircraft, for example, in tailplanes and main rotor blades for helicopters. A modern airliner may have about two tonnes of carbon fibre-epoxy resin composites alone.

Sterilised, tough plastics are used in medicine for tubing, drip feeds, blood bags and components for kidney dialysis machines. Plastics are used in replacement surgery, taking the place of heart valves, hip joints and, of course, false teeth.

There are indeed few facets of modern life where plastics are not used. In many cases they have made possible products that could not have existed without them. In other cases the ability to make things cheaper has enabled many more of us to enjoy a standard of life undreamt of by our forebears. They are such a significant part of life today that it is important that we should try to understand what they are and what they can do. It is hoped that this booklet will help in this task.

Because of their high strength–weight ratio fibre-reinforced plastics are widely used in the aircraft industry. For example, up to 1800 kg of carbon fibre-epoxy resin composites are used in a modern airliner whilst some modern aircraft consist of about 20 per cent, by weight, of such materials. Helicopter rotor blades, such as the sectioned blade illustrated on the left, are made from composite materials based on glass and carbon fibres with thermosetting plastics. Such blades have better fatigue resistance than metal blades.

A ski boot.

The nose cone of Concorde is made from fibre-reinforced plastics.

A Definition of Plastics

The nature of plastics will be explained in some detail in later sections. In the meantime it is useful to try and define what is meant by the word *plastics*. In practice it is surprisingly difficult to give a simple definition which covers those materials people generally consider as plastics, and exclude such materials as glass, metals and rubber. For our purposes, however, plastics may be considered as man-made materials based on very large molecules which at some stage are capable of flow under pressure (and usually heat), enabling them to be shaped into the form required.

2 The First Plastics

Plastics in the ancient world

While plastics are regarded as a product of the twentieth century their origins may be found in antiquity. The fossilised resin amber has been a prized gemstone for thousands of years: specimens have been found at Stonehenge, in Mycenaean tombs and in ancient European lake dwellings. Amber may be considered as a plastics material since fragments resulting from machining operations may be moulded under pressure at about 170°C, a process that was developed commercially in about 1880 to give a product known as *Ambroid.*

Articles made from Parkesine between 1860 and 1868. Developed by Alexander Parkes in North London it was the forerunner of Celluloid and Xylonite.

In Victorian times gutta percha was widely used to produce highly ornate mouldings such as this inkstand. Until the Second World War it was also used for undersea cable insulation and for bottles used to store hydrofluoric acid until replaced in each case by polyethylene. Like natural rubber, to which it is closely related, it is obtained from the latex of trees.

A number of other naturally occurring materials were also known to the ancient world, including shellac. This was used widely in the first half of this century to make gramophone records, whilst another natural material, *bitumen*, has only recently been replaced with a synthetic plastic in the casings of low-cost battery boxes. Gutta percha, closely related to natural rubber, was widely used in Victorian times.

Ebonite, celluloid and related materials

These naturally occurring materials were originally used without any deliberate modification although some change in the chemical structure may have occurred in some cases on exposure to light. Ebonite was the first plastics material which resulted from deliberate modification. This rigid, black, ebony-like material is obtained by heating natural rubber with about half its weight of sulphur. Although the first patent was taken out by Nelson Goodyear in 1851, both Charles Goodyear in the United States and Thomas Hancock in England had produced similar materials during the previous decade.

In 1862 another British inventor, Alexander Parkes, was awarded a bronze medal for his exhibit Parkesine which was displayed at the Great International Exhibition held in London. This material was made by reacting the natural material, cellulose, with nitric acid and blending the resultant cellulose nitrate with certain solvents and with vegetable oil. Parkesine was used to make articles such as combs, umbrella handles, knife handles and chessmen. Some examples of these early plastics products still exist today. Unfortunately for Parkes, his Parkesine company went into liquidation in 1868, but his work paved the way for the development of the closely related material, Celluloid, by the Hyatt brothers in the United States, and of Xylonite by Daniel Spill in England. In essence these materials were blends of cellulose nitrate and camphor, the latter imparting the characteristic odour of these materials (which in recent years have been most commonly used to make table tennis balls).

One of the most serious disadvantages of Celluloid and related materials is its high flammability. There is an interesting story recalled by John Wesley Hyatt concerning one of their products, billiard balls: 'In order to secure strength and beauty only colouring pigments were added, in the least quantity. In effect this meant that the billiard balls were coated with a film of almost uncontaminated guncotton. Consequently a lighted cigar applied would at once result in a serious flame and occasionally the violent contact of the balls would produce a mild explosion like a percussion guncap. We had a letter from a billiard saloon proprietor in Colorado, mentioning this fact and saying that he did not care so much about it but that instantly every man in the room pulled a gun.'

The first synthetic plastics

It was not until the first decade of the twentieth century that the first truly synthetic plastics material was successfully developed. Building on earlier work carried out in Russia, Germany and England, Leo Hendrik Baekeland, (a Belgian-American) reacted phenol and formaldehyde (methanal) under carefully controlled conditions to obtain a hard heat-resisting material which he marketed as Bakelite. Within a few years this trade name had become a household

Both still and cine photography today are based on the use of plastics films as the carrier for the photographic emulsion or magnetic material as in this slide. Cellulose nitrate products such as Celluloid are no longer used for this purpose because of their flammability.

Phenol-formaldehyde plastics (phenolics) were rapidly accepted by the new radio industry. The photograph shows a phenolic radio housing of the 1930s. The unusually light colour (for a phenolic) is the result of a special manufacturing technique.

word and the material became widely used in the infant electronic and car industries. Very similar materials were used for laminated plastics and in plywood manufacture. Even coffins were moulded from this material.

One disadvantage of Bakelite-type materials (known generically as phenolic resins) is that they are normally dark in colour. No such colour restrictions occurred, when materials were made by reacting urea and formaldehyde. These urea-formaldehyde plastics were introduced commercially in 1928. They were available in a wide range of colours, superior to the phenolic materials in many of their electrical insulation properties, but inferior in heat and water resistance. Today they are widely used for electrical plugs and sockets, toilet seats and screw tops for jars, as well as wood adhesives, which are particularly useful in the manufacture of chipboard.

The discovery of polyethylene

The 1930s saw the emergence of many new plastics materials including polyethylene (known originally as polythene), polyvinyl chloride (PVC), polystyrene and polymethyl methacrylate. Their development came about not simply as a result of much careful laboratory experimentation but also because research chemists were careful to record and, later, to consider and exploit unexpected results. The discovery of polyethylene is a good example.

During the early 1930s workers at ICI were carrying out experiments at very high pressure (3,000 atmospheres) trying to bring about reactions which did not occur at normal atmospheric pressure. The results seemed to be generally disappointing. However, the two chemists carrying out the research programme, E.W.

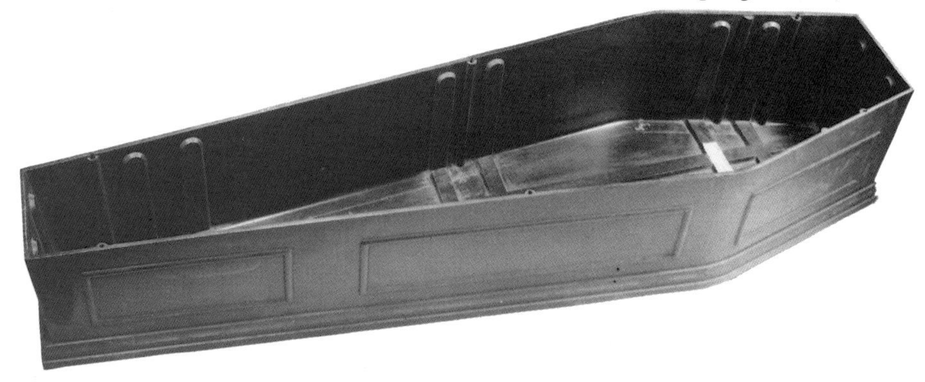

A coffin (with a cut-away lid) made from phenolic plastics in the 1930s.

Fawcett and R.O. Gibson, noted that in some attempted reactions which involved the gas ethylene, tiny white particles were to be seen floating about in the unreacted liquid. In a later experiment, on 27 March 1933, they attempted to react ethylene and benzene at 170°C and 1,400 atmospheres pressure. After the experiment, seemingly without reaction, they noticed a white waxy solid coating the walls of the reaction vessel.

Subsequent analysis showed this to be polyethylene. This was an unexpected result.* The ICI chemists repeated the experiment using ethylene alone and once again a small amount of polyethylene was produced. Subsequent attempts to make larger quantities by reacting at higher pressure led to an explosion which destroyed parts of the laboratory. When, in December 1935, an experiment led to the production of eight grammes of polyethylene it was felt that considerable progress had been made. It was subsequently found that this experiment had only been 'successful' because there had been a small leak in the apparatus which had resulted in exactly the right concentration of oxygen in the apparatus necessary to catalyse the reaction. Slowly the reaction became better understood and on 1 September 1939 – the day Germany invaded Poland and the Second World War broke out – ICI started operation of a pilot plant for the production of polyethylene.

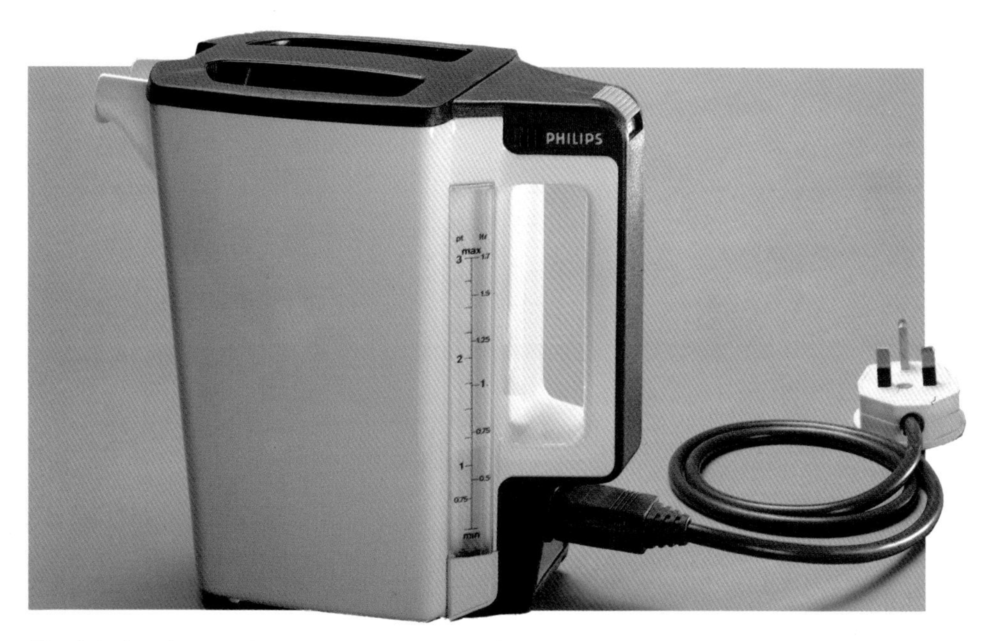

Electric kettle jugs have enjoyed rapid acceptance because of their space saving shape, their capacity to boil only small amounts of water if required, and their design flexibility.

Nylon and other thermoplastics

Whereas the development of polyethylene originated from the observation and following up of an unexpected experimental result, the development of nylon fibres and plastics was more deliberate. In 1928, W.H. Carothers joined the Du Pont company which wished to synthesise a material capable of replacing silk as a fabric in many applications. In his studies Carothers produced many new materials such as polyesters and polycarbonates but it was not until 1935 when he reacted hexamethylene diamine with adipic acid to produce a material now known as nylon 66 that he achieved his target. Commercial production of nylon 66 commenced in October 1938, after Du Pont had spent $27,000,000 (then an enormous sum of money) on its development.

In addition to the successful introduction of a material initially important as a filament and fibre and later as a plastics material, the work of Carothers had wider importance. This was because his work provided the theoretical basis of that branch of chemistry concerned with the synthesis of plastics, rubbers, fibres and related materials, known as polymer chemistry.

*It was unexpected because up until that time it was believed that there were good theoretical reasons why polyethylene could not be made from ethylene.

Plastics as replacements for traditional materials

Before the Second World War most of the newer plastics were little more than curiosities. Polystyrene for example was a very expensive electrical insulator which cost about fifty dollars a pound, then about five times the weekly income of a skilled worker in Britain. War-time needs, in particular the problems arising from the unavailability of natural rubber to most of the combatants, led to considerable developments and by the 1950s polyethylene, PVC and polystyrene had become available as low-cost plastics, whilst polymethyl methacrylate, (probably best known as the sheet material Perspex), had become a familiar material.

Such a ready availability of materials at low cost soon led to serious problems. A lack of understanding of the properties of plastics, poor design and the use of inferior materials led to many bad applications, and plastics acquired a poor reputation. To counter this, the major suppliers of plastics materials initiated a large education exercise, set up huge technical service laboratories and published extensive technical literature. This resulted in a better appreciation of the capabilities and the limitations of the materials available and in consequence generally much improved product design.

Over the past 35 years there have been few new major tonnage plastics,[†] polypropylene being a well-known exception, but the overall market has grown enormously. In part this is because the cost of making plastics materials and shaping them into finished products has increased more slowly than for many more traditional materials in spite of the huge increase in the cost of oil (the main raw material for plastics) in the 1970s. Consequently it has frequently become cheaper to use plastics than to use metals, ceramics and other traditional materials.

The use of plastics as a replacement for metals may be economic for a number of reasons. Firstly, the cost of the raw material to make the product may be less. Secondly, most plastics materials are highly suitable for mass production techniques at temperatures much lower than those used to melt metals, and generate little scrap. Thirdly, it is often possible to make highly complex mouldings in one operation where previously the part had to be made by shaping and then assembling several pieces of metal.

Plastics for special purposes

In addition to the major tonnage materials, more specialised plastics have found important outlets. Due to their toughness, excellent abrasion resistance (superior to many metals) and resistance to hydrocarbon oils, the *nylons* have become well-established for gear wheels, bearings and many light engineering applications. One disadvantage of the nylons is their tendency to absorb water. In this respect the *polyacetals*, also first developed by Du Pont, are somewhat better and although their abrasion resistance is not quite so good they find many similar uses. Polytetrafluoroethylene (PTFE) has become well known in non-stick ovenware as well as being used in many hundreds of industrial applications.

Over the past 20 years several new plastics materials have appeared which show characteristics such as toughness and heat resistance. When properly used, the polycarbonates can be exceptionally tough. This was demonstrated when they were used as face shields by astronauts on their lunar space walks.

The many other uses of these materials include camera bodies, power tool bodies, food processor bowls and audio compact discs. In the past, they have also been used for safety helmets, but many grades are liable to crack in the presence of certain liquids, including some paints. Other plastics which are used in engineering applications include polysulphones, polyimides, polyphenylene oxides and polyphenylene sulphides.

As long ago as 1942 Whinfield and Dixon, when working in England for the Calico Printers Association, dis-

[†]Any definition of a major tonnage material must be arbitrary but for this booklet we will take it to mean plastics which are produced in quantities in excess of 1,000,000 tonnes per year.

covered polyethylene terephthalate. This became very well known within a few years as the fibres Terylene and Dacron. It also became available in the form of film such as Mylar and Melinex. In more recent times, methods have been devised to make bottles and other containers out of this material. The bottles have been

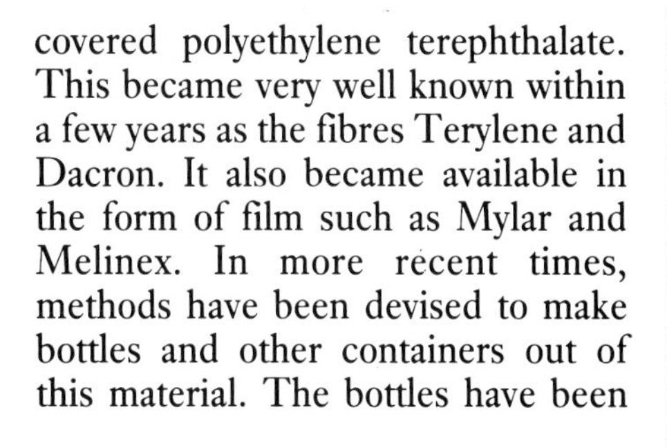

Polycarbonates are widely used for compact discs.

Bottles of all shapes and sizes are now made from plastics. Plastics bottles are particularly valuable in the bathroom since unlike glass they are not liable to break on dropping onto hard surfaces. Polyethylene terephthalate bottles (made from the same polymer as Terylene and Dacron fibres) are widely used in the marketing of carbonated drinks and beers. They have adequately low gas permeability and are much lighter in weight than those made of glass, when empty.

Polyester-glass fibre composites are widely used in sports car bodies. They are strong, light in weight and may be made with less expensive tooling than is used for mass-produced metal car bodies.

particularly successful for marketing beers, colas and other carbonated drinks due to their low gas permeability and lightness compared with glass.

Polyethylene terephthalate is an example of a polyester of which there are many types. Some types show many of the characteristics of liquid crystals when molten. Known as liquid crystal polymers, the solidified materials have exceptional strength, toughness and stiffness, and have only appeared recently on the market.

Rather similar are certain polyamides, chemically related to the nylons. One of these was used to produce an extremely strong fibre, marketed by Du Pont as Kevlar, which is now widely used as a tyre cord fabric.

While most plastics have very good resistance to water some plastics swell or even dissolve in it. One particular type of plastics material, chemically similar to Perspex which swells extensively in water, is used to make soft contact lenses.

One disadvantage of plastics is that although they are lighter in weight, volume for volume, they are usually much less stiff and less strong than metals. It has long been recognised that the combination of the plastics material with a fibrous product could be beneficial. The resultant products have become known as composites. Paper-based laminates such as Formica have become well-known in the home. In the 1950s, glass-fibre rein-

Lorry cabins may also be made from polyester-glass fibre composites.

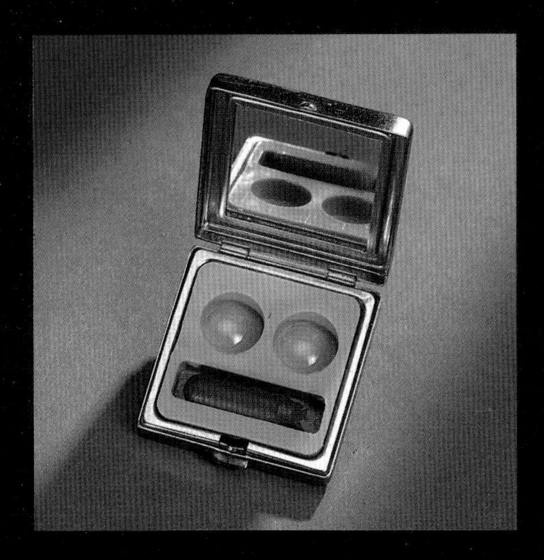

This pair of acrylic micro-corneal contact lenses dates from the 1960s.

forced plastics began to be used for sports car bodies, boat hulls and for a variety of industrial applications. The discovery of carbon fibre led to the production of even stronger composites; these were not only used in aerospace applications but also for tennis racquets. In the late 1980s composites based on carbon fibre and Kevlar have become important structural components of helicopters.

Carbon-fibre reinforced plastics are used for tennis racket frames.

Recently, highly specialised plastics such as polyether ether ketones and polyetherimides have been used with carbon fibres to make composites which possess exceptional strength and heat resistance.

Many newer plastics have unusual properties which still need to be exploited fully. Polyvinylidene fluoride exhibits piezoelectricity – that is, when a slab of the material is subjected to pressure, charges of positive and negative electricity are produced on opposite faces. This property has been of interest to telephone manufacturers and utilised in the making of earphones for personal stereo equipment. Polyvinyl carbazole may be made photoconductive, that is, it becomes electrically conductive in the presence of light. This has been made use of in electrostatic dry copying machines. A polydiacetylene derivative has magnetic characteristics and it is possible to make the needle of a compass from this material without any metal being present. Whilst many plastics have been recognised for many years to be excellent electrical insulators there are high hopes that some of the newer materials will become important as conductors. Each year the range of plastics properties continues to grow. This will lead to an ever widening number of applications which, if used properly, should help us to improve the quality of our lives.

3 Plastics and Polymers

Atoms and molecules

You will have noticed that many plastics materials have names which begin with the prefix *poly* such as polyethylene, polypropylene, polystyrene and polyvinyl chloride. This is because plastics belong to a class of chemicals known as *polymers*. Before we can understand what this means, we first need an explanation of 'atoms' and 'molecules' and related terms.

If you were to take a glassful of water and drink half the contents, what is left in the glass would still be water. This division could be repeated many times although in practice you would soon be dealing with minute quantities of water, and it would be very difficult to drink exactly half each time. If however we assume that you have the perfect ability to drink just half each time, the point would eventually be reached where if you tried to subdivide the material it would no longer be water. At this stage we would have one *molecule* of water. If this molecule were divided up we would find that it consisted of three *atoms*, two of hydrogen and one of oxygen.

About 90 types of atoms occur in nature and some others have been created in the laboratory. These different types are known as *elements*. Well-known elements include carbon, hydrogen, oxygen, nitrogen, sulphur, iron, copper, gold, sodium and phosphorus. The lightest is hydrogen, an atom of which weighs about 0.000 000 000 000 000 000 000 002 grammes (about a million million million millionth of a gramme). Some idea of just how small this is is obtained if we realise that the weight of a hydrogen atom compared to that of a large spoonful of sugar is in about the same ratio as the weight of that spoonful of sugar to the weight of the earth. When we wish to talk about the weight of the atoms of the other elements it is convenient to express them in terms relative to the hydrogen atom and we call this the *relative atomic mass*. For carbon this is about 12, for nitrogen 14 and oxygen 16. Since a molecule of water consists of two hydrogen and one oxygen atom its mass will be about 18 times that of hydrogen and we say that the *relative molecular mass* or *molecular weight* of water is 18.

Big molecules and polymers

Most of the chemicals encountered in school science syllabuses have relative molecular masses of less than 300. For example common salt (sodium chloride) is about 58, carbon dioxide 44, ethyl alcohol 46 and benzene 78. In contrast the molecules from which plastics are made are very much bigger. For example the average relative molecular mass of commercial polystyrene is about 200,000 whilst that of an acrylic sheet material such as Perspex is of the order of 1,000,000. Whilst the molecule is still small (it will weigh only about a million million millionth of a gramme), it is a giant molecule compared with that of water and this will greatly influence its properties.

Giant molecules occur in nature. Examples include cellulose (present in all plants, for example, cotton is almost pure cellulose), starch, proteins (such as those of wool, silk and leather) and natural rubber. It is worth noting that such natural materials include fibres, rubbers and adhesives. The synthetic fibres, rubbers and adhesives and also surface coatings are also, like plastics, giant molecules of the type known as polymers.

The prefix *poly* means 'many' whilst a *mer* is a repeating unit so that 'polymer' means many repeating units. In practice synthetic polymers are made by joining together small molecules known as *monomers*, the process being known as *polymerisation*. Thus polystyrene is prepared by the polymerisation of styrene, and polyvinyl chloride by the polymerisation of vinyl chloride. The polymerisation process and the types of products that may be obtained from them are discussed in further detail in the rest of this section.

Double bond polymerisation: the polymerisation of ethylene

Ethylene (ethene) molecules consist of two carbon atoms linked together by a double bond and four hydrogen atoms, two of which are attached to each carbon atom. This satisfies the requirement that there should be four links, or valencies, to each carbon and one to each hydrogen. Appropriate treatment of the ethylene molecules causes one of the two links between the carbon atoms to break; this enables the resultant 'free arms' to link to similar molecules. If we represent the carbon atom by a letter C and a hydrogen atom by H the process may be schematically outlined as follows:

$$\begin{array}{ccccccccccccccccccccccccccccc}
H & & H & & & H & & H & & H & & H & & H & & H & & H & & H & & H & & H & & H & & H & \\
| & & | & & & | & & | & & | & & | & & | & & | & & | & & | & & | & & | & & | & & | & \\
C & = & C & \rightarrow & - & C & - & C & - & C & - & C & - & C & - & C & - & C & - & C & - & C & - & C & - & C & - & C & - \\
| & & | & & & | & & | & & | & & | & & | & & | & & | & & | & & | & & | & & | & & | & \\
H & & H & & & H & & H & & H & & H & & H & & H & & H & & H & & H & & H & & H & & H &
\end{array}$$

Ethylene **Polyethylene**

Plastics tubing is widely used in surgery and post-operative care. Shown here are tracheotomy tubes.

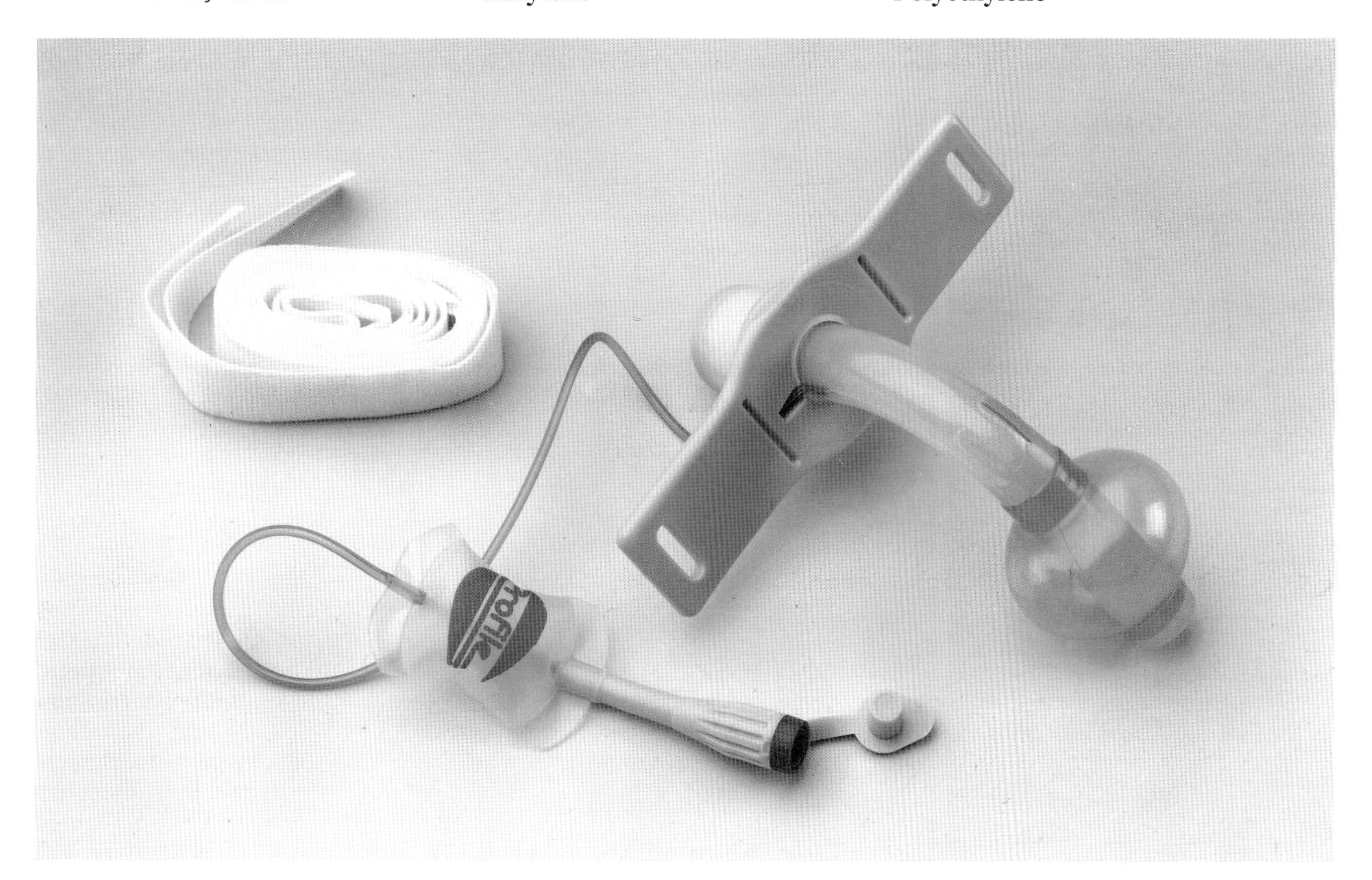

In this way many hundreds or even thousands of ethylene molecules may be joined together in a single chain. This is the process known as polymerisation. The starting material, ethylene, is known as a monomer and the product, polyethylene, is known as a polymer. Since the polymer is obtained by adding the monomers together, this particular process may be referred to as *additional polymerisation* or since it involved the opening of a double bond, as *double bond polymerisation*.

Replacement joints, blood vessels and ligaments.

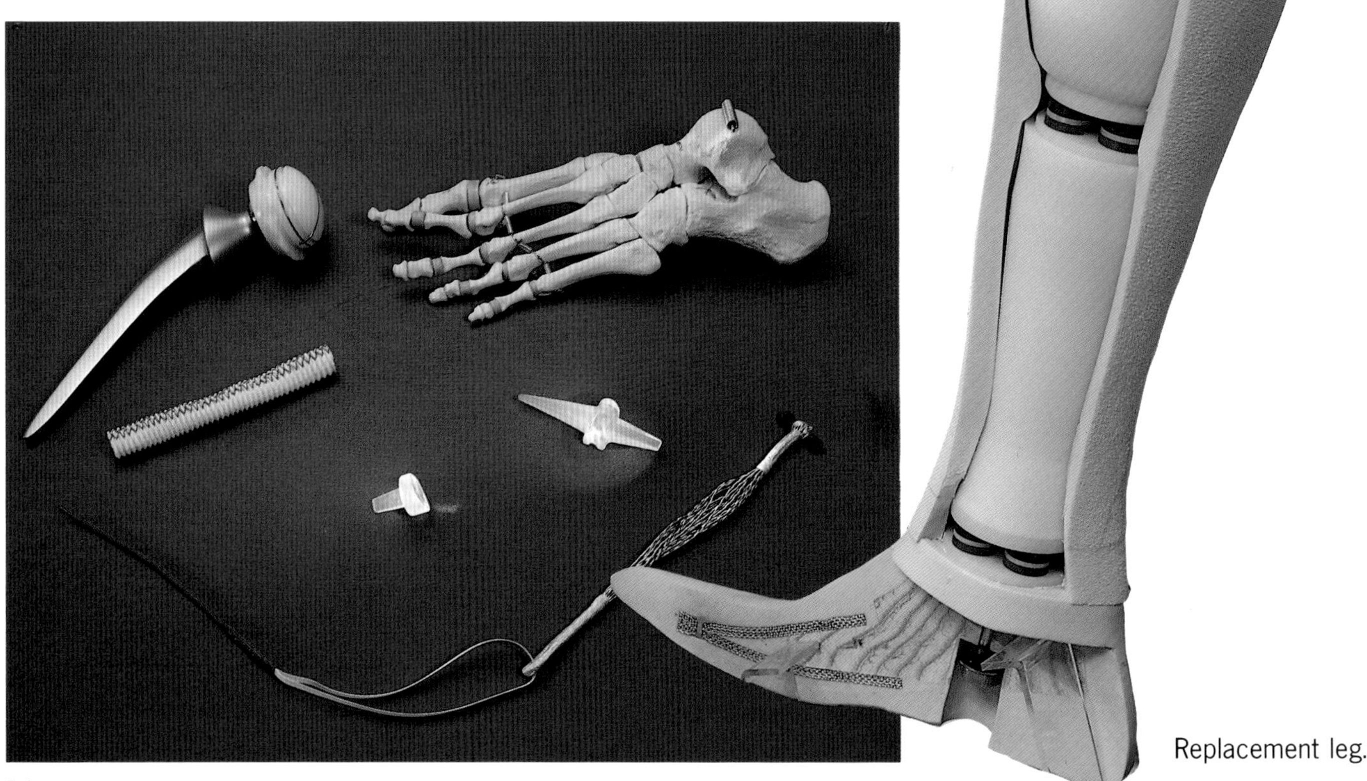

Replacement leg.

Copolymerisation

In many instances it is possible to put two monomers into the reacting vessel and obtain a molecule where both monomers are incorporated into the same chain. Thus if we represented two monomers as A and B the chain might have the following structure:

~AABABBBABAAABABBBAA-BAABABBBABAABBBAABB-BABABBBB~

Such a structure would be called a *random copolymer* (strictly a random binary copolymer as two monomers only were used) whereas we could use the word *homopolymer* to describe a polymer such as polyethylene made from only one monomer. A structure like:

~BABABABABABABABABAB~

we would call an *alternating copolymer* whilst the following would be referred to as a *block copolymer*:

~AAAAAAAAABBBBBBBB~

If we had three monomers then the polymer would be known as a ternary copolymer, usually abbreviated to *terpolymer*. The so-called ABS plastics that are widely used in car interiors and the housings of domestic equipment such as vacuum cleaners and food mixers are terpolymers made by polymerising acrylonitrile, butadiene and styrene.

Ten different plastics materials are used in this domestic food processor: ABS, polyethylene, polypropylene, polyester, polyacetal, nylon, styrene-acrylonitrile copolymer, polycarbonate polymethyl methacrylate, PVC.

Ring-opening and condensation polymerisation

All of the synthetic plastics materials are made by polymerisation processes, as are synthetic fibres, rubbers, surface coatings and adhesives. However, not all polymerisations involve double bonds; for instance, they may involve the opening of a small ring-shaped molecule. An example is the opening of the ring-shaped molecules of caprolactam which then join together to give nylon 6, a polymer important both as a fibre and a plastics material:

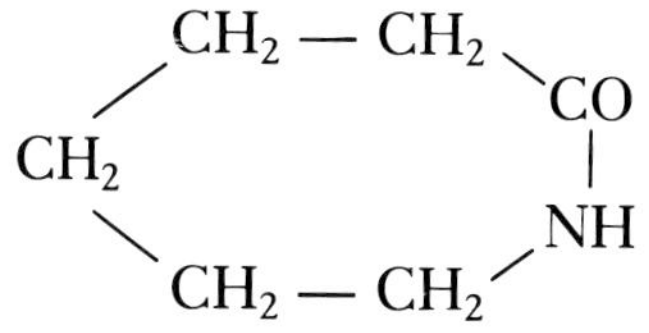

$$\begin{array}{ccccc} & CH_2 & - & CH_2 & \\ / & & & & \backslash \\ CH_2 & & & & CO \\ & & & & | \\ \backslash & & & & NH \\ & CH_2 & - & CH_2 & / \end{array}$$

Caprolactam

$$\downarrow$$

$$\sim (CH_2)_5\,CONH \sim$$

Nylon 6

(In these formulae 0 represents an oxygen atom and N one of nitrogen.)

In addition to being made by ring-opening polymerisation the same polymer may be made by condensation polymerisation of an amino acid, namely amino caproic acid. In this case the amino (NH_2) groups of one molecule react with the acid (COOH) group of another molecule. Eventually a long chain molecule is formed with a molecule of water being formed (and split out or condensed out – hence the name of the process) each time the amino and acid groups react.

$$H_2N.CH_2.CH_2.CH_2.CH_2.CH_2.COOH$$
$$\downarrow$$
$$\sim NH.CH_2.CH_2.CH_2.CH_2.CH_2.CO \sim + H_2O$$

Formation of Nylon 6 from ω-amino caproic acid

Thermoplastics and thermosetting plastics

Polymers such as polyethylene, polypropylene, polystyrene and ABS are known as thermoplastics. These soften on heating so that they may be shaped by moulding and extrusion processes, but they then harden on cooling. This process may be repeated so that scrap mouldings may be reprocessed (although in practice there is a limit to the number of times that this reprocessing may be carried out because repeated heating of the material may cause it to break down).

Such materials are thermoplastic because they consist of long chain-like molecules. At normal room temperatures these molecules may be considered as partially or totally frozen and immobile. On heating they begin to move in a manner reminiscent of the wriggling of worms and as they do so they tend to push each other apart. On application of an external force the molecules are able to slip past one another and the mass flows.

There are however a number of plastics materials that at the end of processing operations are not simply long chains but three-dimensional networks and these will not be able to flow on heating. In practice these networks are often formed by the joining of small molecules in three dimensions (cross-linking) during a heating process, that is to say that the shape is set by heating; such materials are known as thermosetting plastics. Important thermosetting plastics include the phenolic resins and urea-formaldehyde resins mentioned in the previous chapter;[†] the polyester resins used for making glass-reinforced laminated boats, canoes and car bodies; and the epoxy resins which are best known as adhesives.

These two fundamental types of plastics materials are illustrated in the diagram:

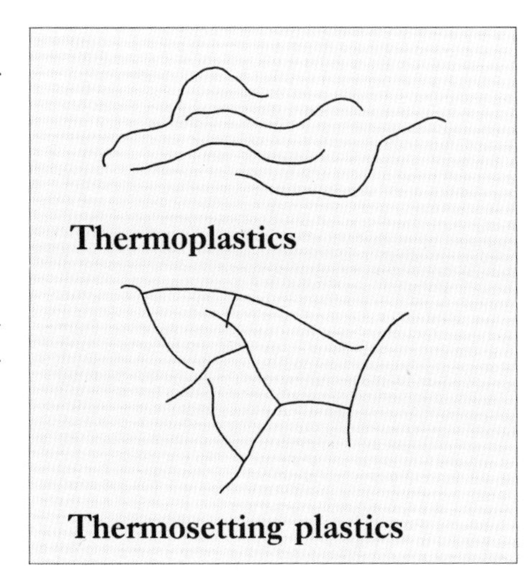

The thermoplastics, in turn, may be subdivided into two groups: *amorphous thermoplastics* and *crystalline thermoplastics*. In the case of amorphous thermoplastics, such as polystyrene and ABS, a lump of the material would comprise millions of irregularly curled-up molecules. Such materials tend to be glass-like and in the absence of additives and air bubbles, transparent. Unless modified by additives they are also generally rigid and are rather brittle.

Thermoplastics such as polypropylene and the nylons are said to be crystalline. This does not mean that a piece of nylon has a definite external shape such as possessed by a grain of common salt or a crystal of copper sulphate. However, within the total mass, parts of the molecules may pack together in a somewhat orderly fashion. As a result, only part of the mass of material will be in crystalline form and mixed up with amorphous material. In the case of polyethylene the amorphous part is quite flexible at room temperature and the crystalline zones effectively hold it together.

[†]Although methods are now available to cross-link and harden many of these materials at room temperature, the term 'thermosetting material' is still retained.

By varying both the polymerisation method and the shaping operation, products may be obtained with different proportions of crystallinity and, consequently, different levels of stiffness.

Plastics materials are seldom simply polymers. Additives such as pigments, antioxidants and fillers are almost always present. In a number of applications the polymer may be mixed with glass fibres or carbon fibres to give products of increased strength and stiffness, the products being known as *polymer composites*. A number of important plastics consist of polymer blends; blends of polymers which have properties not shown by single polymers.

Summary

At this stage let us summarise what has been said on the nature of plastics:

i) Plastics, rubbers, fibres and synthetic adhesives are made from materials consisting of very big molecules.

ii) Some big molecules such as cellulose and the proteins occur in nature. When, however, it is desired to make big molecules synthetically this is achieved by using the process of polymerisation on simple materials known as monomers.

iii) Copolymers are produced where more than one monomer is used.

iv) There are two main types of plastics: thermoplastics and thermosetting plastics.

v) The thermoplastics may be amorphous or crystalline. They may also be blends of polymers or blends with glass or carbon fibre.

In the next section we will see how they are processed to produce useful articles.

4 The Shaping of Plastics

In principle the shaping of plastics materials involves two stages which may be summed up in the two phrases:

Get the shape
Set the shape

Getting the shape usually involves melting the material and shaping by processes such as moulding, extrusion and calendering. In the case of thermoplastics, *setting* is usually brought about simply by cooling, whereas with thermosetting plastics chemical reactions occur which cause the material to cross-link into a rigid network.

Compression moulding

The first process to be developed on a large scale and still of some importance is that of *compression moulding* (figure 1). In this process the material is heated in a mould and then pressed into shape by means of a modern version of the hydraulic press (invented by the English engineer Joseph Bramah in 1795). It is most suitable for thermosetting plastics because the setting reaction takes place in the mould and the moulding may be taken out whilst hot. In the case of thermoplastics the mould first has to be heated to soften the plastics material in order to shape it and it then has to be cooled in order to remove the moulding from the mould. Probably the most well-known examples of compression mouldings are domestic plugs and switches (normally made from urea-formaldehyde plastics).

Injection moulding

The process of injection moulding is of much greater importance today. Whilst rudimentary injection moulding equipment was first used by the Hyatt Brothers, the inventors of Celluloid, the modern injection moulding machine originated with the work of the Germans Eichengrün and Buchholz in 1921, and the first commercial production machines of Eckert and Ziegler in 1926.

Injection moulding consists of melting the polymer in a cylinder and then injecting it into a mould where the material sets. In the case of thermoplastics the mould is below the setting temperature of the plastics material used, whilst with thermosets the mould is sufficiently hot to cause the material to harden by cross-linking.

Modern injection moulding machines come in all sizes. The smaller machines are only able to mould about 20 grammes at a time, and the largest about 60 kilogrammes. It is therefore possible to mould a huge range of products ranging from tiny clips, gear wheels, combs, computer keyboard keys and other small items to large dustbins and small boat hulls. The word processor on which I am preparing this text has a keyboard, monitor and printer, the housings for which, as well as many other parts, are injection moulded.

Early injection moulding machines suffered from the problem that it was

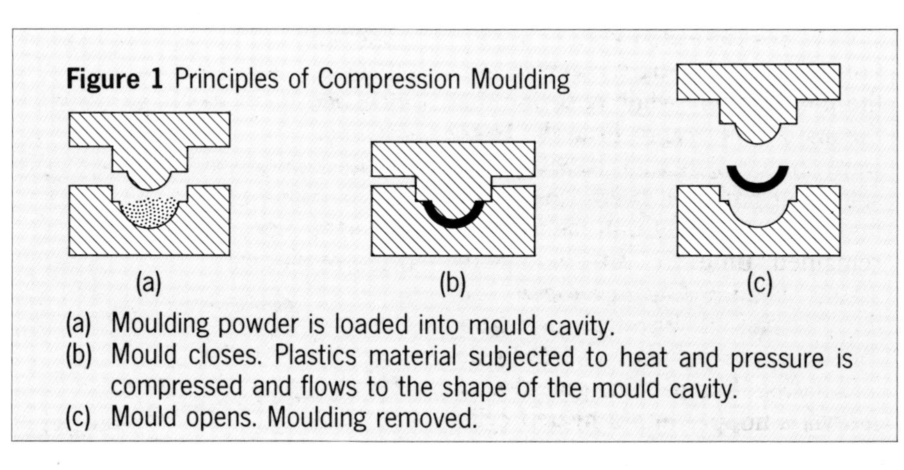

Figure 1 Principles of Compression Moulding

(a) Moulding powder is loaded into mould cavity.
(b) Mould closes. Plastics material subjected to heat and pressure is compressed and flows to the shape of the mould cavity.
(c) Mould opens. Moulding removed.

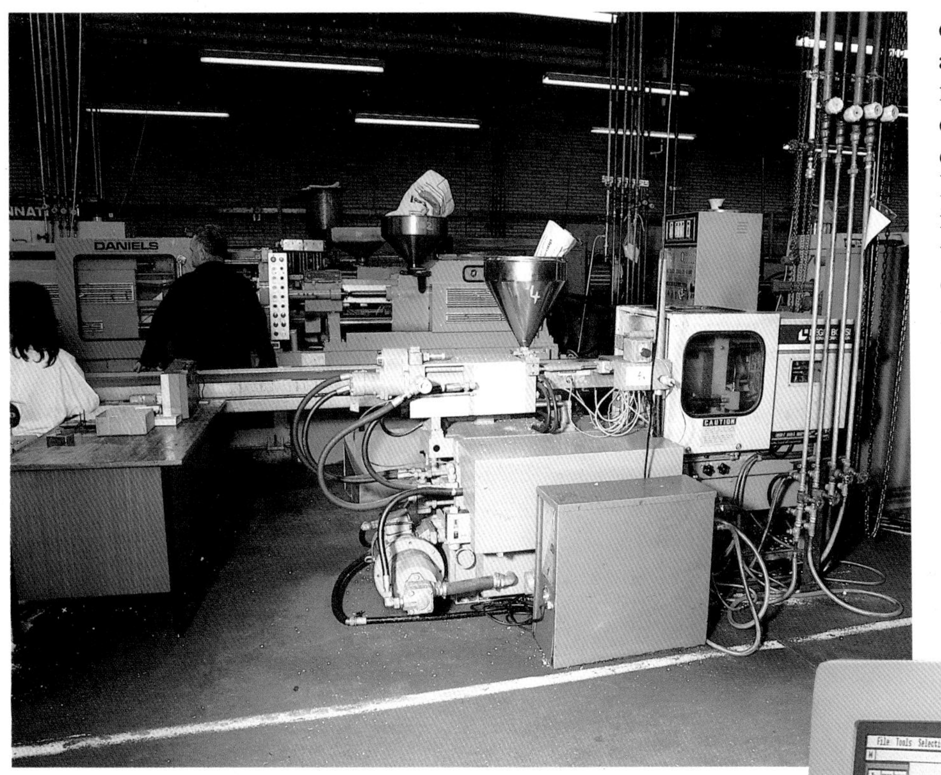

A modern injection moulding machine.

difficult to achieve a good transmission of pressure between the injection plunger and the melt flowing in the mould cavity while ensuring that the melt was evenly heated. There were many ingenious attempts to solve this problem which was affecting product quality but the problem remained until the in-line single-screw machine was developed. Today nearly all injection moulding machines are of this type, and work as follows. The plastics granules are fed via a hopper into a heated cylinder. At the same time the granules are heated and melted (both by friction and externally applied heat), compacted together, and the melt is churned around to aid even heating. Rather than pump the melt straight into the mould it is usually preferable to collect the melt at the front of the cylinder. In order to do this the screw moves backwards at the same time that it is rotating until there is enough melt at the head of the machine. The screw then stops rotating and moving forward like a ram it injects the melt rapidly into the mould. While this melt is cooling in the mould cavity

Plastics are widely used for housings of electronic equipment. The main requirements are rigidity, toughness, good appearance and good electrical insulation characteristics. The keyboard, visual display unit and printer of the word processor are made of plastics.

the screw starts to feed more melt into the head of the cylinder ready for the next 'shot' (see figure 2).

Whilst in principle the process is simple, successful moulding requires attention to choice and care of material, selection of moulding conditions, product design and design of the mould. With modern equipment it is possible to operate under very carefully controlled conditions of temperature, pressure and time which may be monitored both on visual display units or recorded. Optimum setting conditions for a given mould and plastics material may be recorded, for example, on tape or on a programme card. This may then be inserted into the machine control system on a future occasion where that particular combination of mould and material are to be used, and the system automatically sets the machine working to these conditions. Under such close control it is possible to achieve a high degree of consistency even with the most complex mouldings.

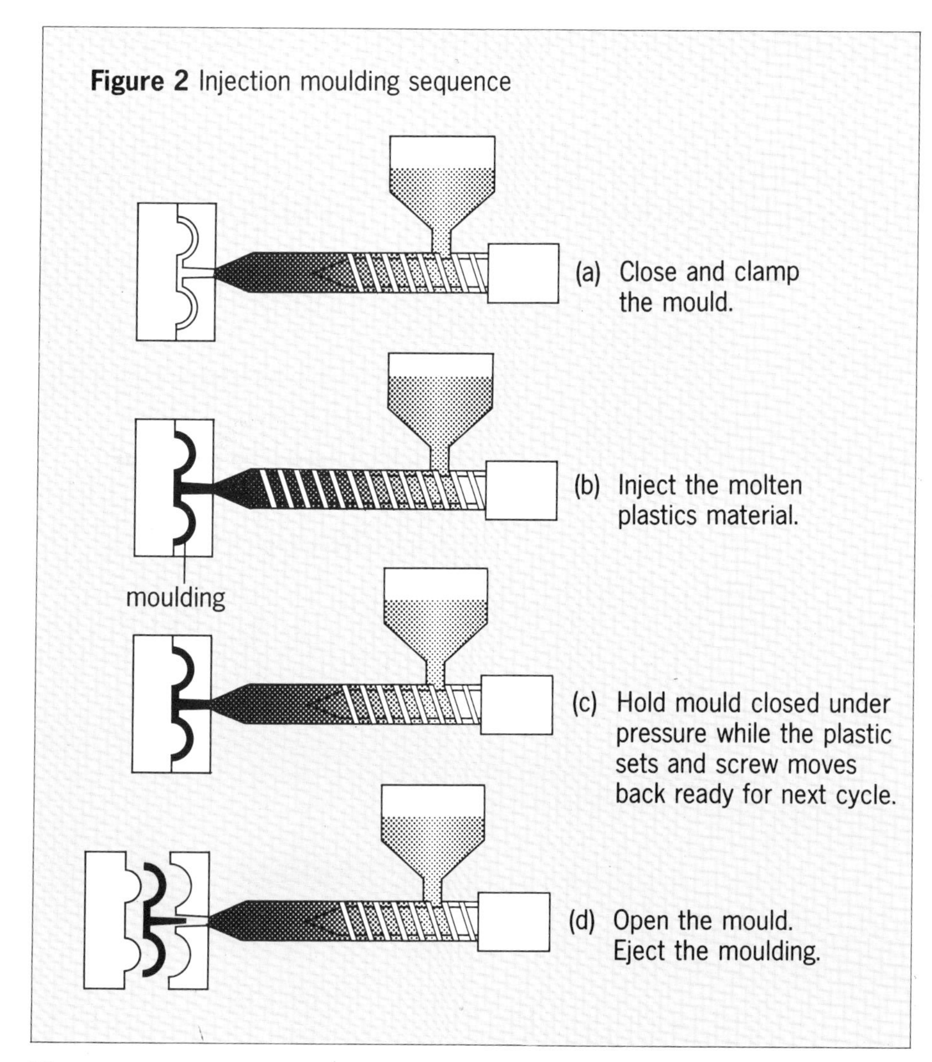

Figure 2 Injection moulding sequence

High levels of automation are possible with automatic loading of hoppers and robots to remove the mouldings once they have been made. Working on a 10-second cycle and assuming that the machine is working 24 hours a day it would take less than six days to make 50,000 mouldings. With some moulds it is possible to make several mouldings or 'impressions' per cycle so that under the above conditions a 20-impression mould could make a million mouldings within the same amount of time.

Extrusion

Where the product has a constant cross-section, the extrusion process may be used. As with injection moulding the granules are fed via a hopper to a heated cylinder or barrel (the latter term usually being used here). The screw rotates (but does not reciprocate as in injection moulding) and pumps the melt through a hole or gap in a die. Variations of the extrusion process are used to make such differing products as ball-point pen ink tubes, polyethylene bags, plasticised PVC garden hose, sheet for refrigerator linings and coverings for wire and cable. Since extrusion is also used in many of the processes involved in the manufacture of plastics materials, including blending and colouring, it is probably true that a greater volume of plastics materials are subjected to extrusion operations than to any other process.

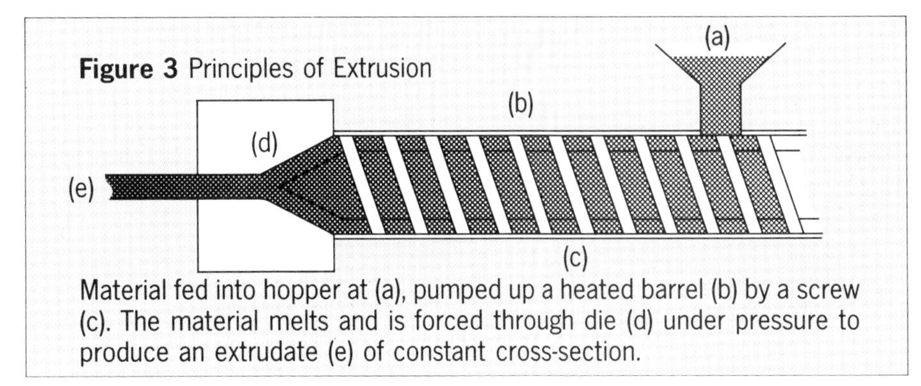

Figure 3 Principles of Extrusion

Material fed into hopper at (a), pumped up a heated barrel (b) by a screw (c). The material melts and is forced through die (d) under pressure to produce an extrudate (e) of constant cross-section.

As with injection moulding the process in principle is very simple (see figure 3). Indeed it has often been likened to the process of decorating a cake with an icing gun. In practice the preparation of extrudates (i.e., the products of extrusion) of good quality and consistent dimensions once again requires the use of equipment operating under closely controlled conditions. One interesting variation in the process occurs where it is required to produce a film comprising layers of different materials – for example, one layer may reduce the permeability to gases whilst the layer on the opposite side of the film may have good resistance to abrasion. Such multilayer films may be obtained by co-extrusion from two or more extruders into a single die.

Part of an extrusion line used to produce co-extruded film (BXL Flexible Packaging Division).

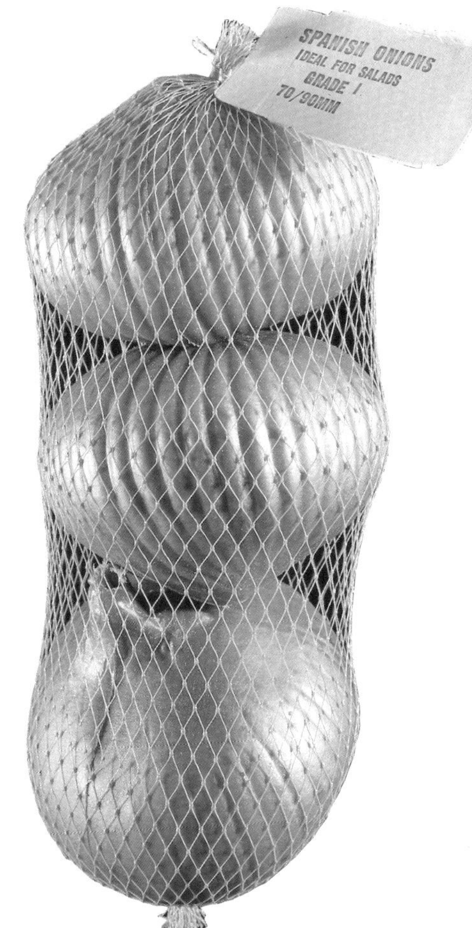

Netlon is made by an extrusion process in which two die parts rotate in opposite directions. Each die part has slots through which strands of molten polymer emerge. When the slots meet, the two strands fuse together to form a network structure. This photograph shows onions wrapped in Netlon.

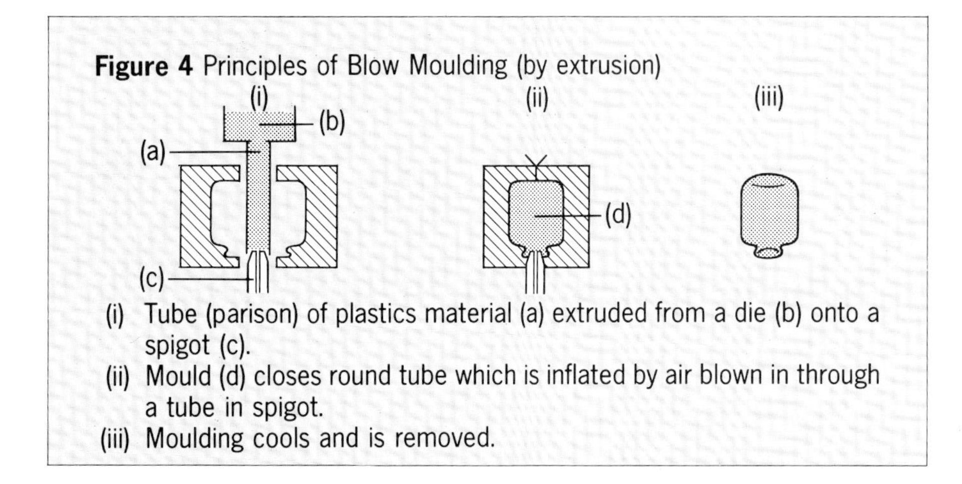

Figure 4 Principles of Blow Moulding (by extrusion)

(i) Tube (parison) of plastics material (a) extruded from a die (b) onto a spigot (c).
(ii) Mould (d) closes round tube which is inflated by air blown in through a tube in spigot.
(iii) Moulding cools and is removed.

Blow moulding

Blow moulding is a process that lies to some degree between injection moulding and extrusion. There are many variations on this process but the most well-known is that in which a tube of polymer melt is extruded downwards out of a die (figure 4). A mould then closes around a portion of the tube known as a *parison*, air is blown into it and the tube inflates to the shape of the mould cavity. The process is particularly useful for making hollow objects such as bottles and containers.

One common problem which occurs in blow moulding is that when a tube is inflated into the shape of a bottle the corners of the bottle tend to be rather thinner than the side walls since the tube at these corners has to be stretched rather more than elsewhere. It is possible to vary the thickness of the tube (parison) wall during the extrusion, by a process known as parison control, so that the thickest section of tube corresponds to the part of the tube that has to be inflated the most.

Rotational moulding

Another process capable of making large, often hollow objects, is that of *rotational moulding*. In this process granules of polyethylene are charged into a hollow mould which is then closed and rotated about two axes in a large oven. The polyethylene melts and coats the inside mould surface. The mould is then removed from the oven and cooled while continuing the rotation. Both the body and lid of the car trailer shown here were made by rotational moulding of polyethylene.

A car trailer, rotationally moulded from high density polyethylene. The design has good aerodynamics reducing drag whilst the solid lid is both lockable and protects the contents from the weather.

Amongst the advantages of this design are that the contents are well protected from the weather and the lid is lockable.

A variation of the rotational moulding process is useful for making balls. In this case a mixture of PVC powder in a liquid plasticiser is formed into a paste which is put into the mould. Once again the paste coats the surface but when this is heated it gels to form a rubbery coating. As before, the mould is withdrawn from the oven and cooled before the mouldings are extracted. This is an example of a process which does not require melting of the polymer to get the shape.

Calendering

Whilst plastics sheeting is often made by extrusion, it may also be made by calendering, a process used particularly with plasticised PVC. Calendering consists of squeezing or flattening softened material between hot rollers to give a smooth sheet. In some ways this is reminiscent of the operation of a rotary iron, and calenders have also been used for many years to impart a finish to cloth.

Other shaping processes

All of the processes described above, with the exception of rotational moulding of PVC pastes, require that the polymer is melted. This is not the case with *vacuum forming*. In this process a thermoplastics sheet is clamped over a box containing a

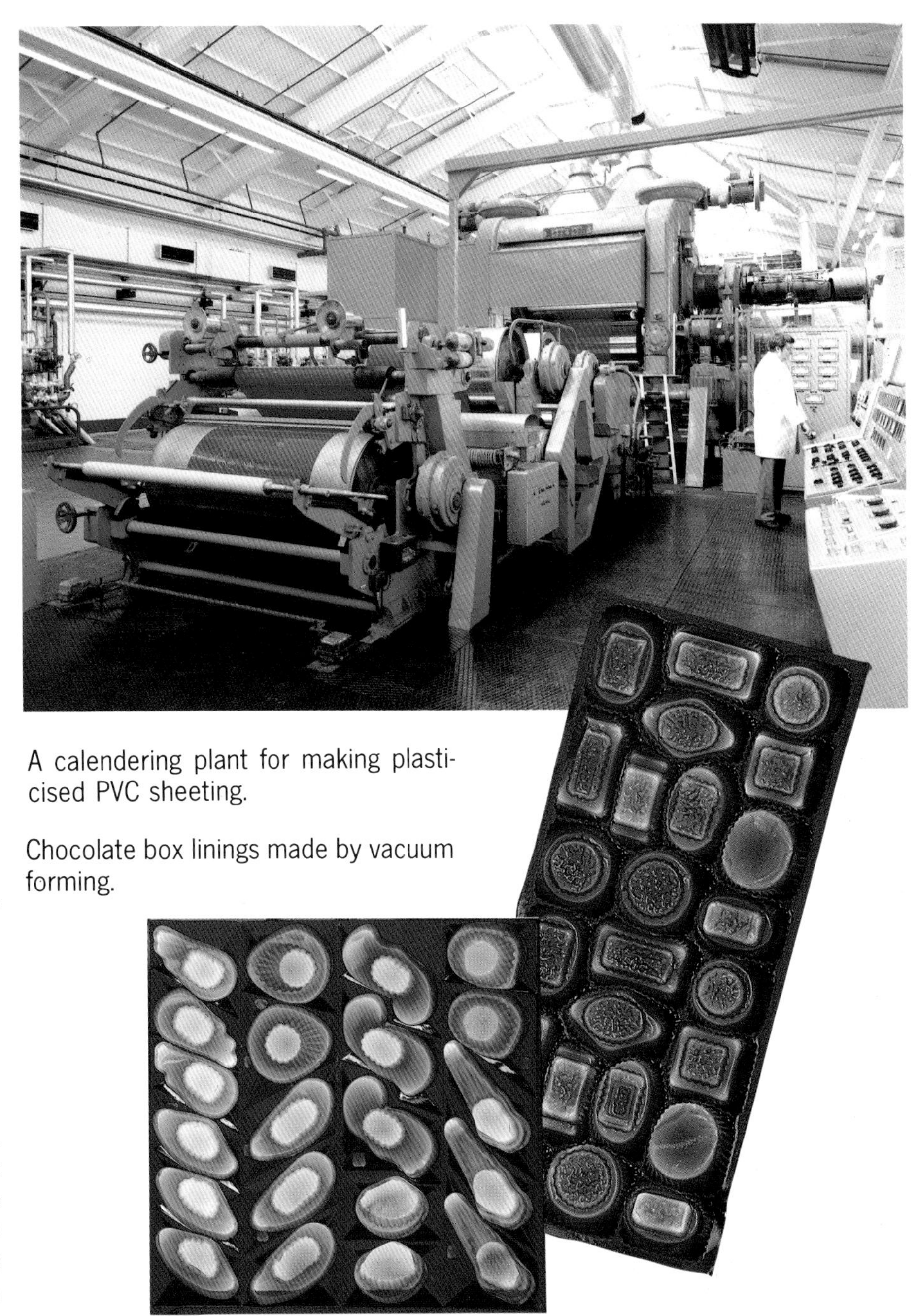

A calendering plant for making plasticised PVC sheeting.

Chocolate box linings made by vacuum forming.

mould or 'former'. The sheet is heated up and the air inside the box is removed by applying a vacuum so that the sheet takes up the shape of the former. This process is widely used for making such diverse products as refrigerator liners, display signs, chocolate box trays, disposable drinking cups and the sides of caravans. It is limited by the fact that the maximum pressure available for shaping is simply atmospheric pressure, and this pressure is not always sufficient. Therefore, for example, in the manufacture of baths from acrylic sheet, it may be necessary to use a compressed air system rather than to rely on atmospheric pressure alone.

Another process that does not rely on the use of melts is that used for the manufacture of *polyurethane foams*. In this process a polymer of low relative molecular mass (<2,500), (i.e., low for a polymer), usually a polyester or a polyether, is reacted with a chemical known as a di-isocyanate. As the materials react, the polymer molecules join together to form a large cross-linked structure. Depending on the choice of raw materials the amount of cross-linking may be light or it may be intense and a spectrum of materials can be obtained which at one extreme are hard and rigid and at the other, soft and rubbery. If water is present it will react with some of the isocyanate present to produce carbon dioxide, a gas, which causes the material to foam. Providing the cross-linking and foaming rates are similar, the finished product will also be a foam. (In practice today foaming is also brought about by the presence of a volatile liquid which vaporises due to the rise in temperature of the reacting mixture during cross-linking.) Rigid foams, such as those widely used in thermal insulation, are produced where there are high levels of cross-linking; moderate levels of cross-linking give the so-called semi-rigid foams (used for car crash pads above the fascia panel); whilst lightly cross-linked materials yield rubbery foams (commonly used in upholstery and for car sponges).

Whilst the above mentioned processes are, in tonnage terms, the most important, there are many others. For example, plastics in the form of a paste or a viscous solution may be spread onto cloth, solutions may be

cast into film, and foams made by beating a polymer latex. In addition there is a variety of processes for making glass fibre-reinforced and carbon fibre-reinforced plastics. Not only do plastics technologists have a huge choice of materials but they also have at their disposal a comprehensive range of processes which must be the envy of those working with more traditional materials.

The front section of this British Rail HS 125 diesel locomotive has a reinforced plastics body.

5 Plastics and the Environment

One frequent result of societies becoming more affluent is that they have more material of which they wish to dispose. Where such waste disposal is not controlled litter will result. When the litter takes the form of broken glass or sharp-edged rusty metals it can be a health hazard. Plastics litter is not usually dangerous but, being brightly coloured, it is usually very visible. In addition, because of its low density it is often washed up on beaches where it becomes an unsightly nuisance; this is aggravated by the fact that plastics do not degrade readily. People cause litter, and in those societies and communities which have resisted the trends to greater slovenliness the problem of plastics litter is much less.

Notwithstanding these comments, there is clearly more plastics litter lying around than any of us want. One approach has been to incorporate into plastics film and sheeting additives which, when exposed to light or to micro-organisms in the soil, will cause the plastics to degrade and to be absorbed into the soil. In at least one country laws have been passed that require plastics films to be biodegradable.

Since much of the litter originates from disposable packaging it is sometimes tempting to suggest the prohibition of such products. This is not however realistic as may be demonstrated by one everyday product – the yoghurt pot. Early re-usable glass containers weighed about 200g; a modern disposable polystyrene pot weighs about 6g. Not only is the pot much cheaper but there are huge savings in transportation costs; furthermore the polystyrene pot does not have to be transported back to the dairy and washed before re-use.

Plastics are products of the chemical industry and this can lead to problems related to the environment. Many of the chemicals used in the manufacture of plastics are reactive, and hence tend to be biologically active and often toxic. When accidents occur which involve these plastics chemicals (for example, collisions involving road tankers), this has led to criticism about plastics.

Fire and flammability are further areas of concern. Many plastics burn, some emitting highly toxic gases and copious amounts of smoke which endanger life. The cost of fire damage can also be very considerable. In recent years much effort has been put into the development of fire-retarding materials, and substantial progress has been made.

Recycling and incineration

The huge increases in energy prices in the 1970s led to extensive interest in the recycling of waste materials, including plastics. Each year many millions of tonnes of waste are collected, the bulk of which are disposed of by land-fill. However, most waste is combustible. The combustible materials are capable of providing a source of energy and it is interesting to note that the calorific value of polyethylene and polystyrene is about three times that of wood. Waste incineration is an important method of waste disposal in a number of countries, and energy is recovered and recycled back into the community. It should be noted that with both land-fill and incineration methods care has to be taken with toxic products. For example, the incineration of PVC leads to the production of hydrochloric acid, although the quantity is somewhat less than half that produced in a typical incineration plant, and precautions must be taken to prevent this acid from polluting the atmosphere.

Although incineration provides energy, plastics are irretrievably lost as raw materials. In some cases it is possible to recycle used plastics waste. Such recycling has been long established in many plastics processing factories where faulty injection mouldings, extrudates and other scrap can be reground and re-used. There are also a very large number of plastics items such as agricultural film, bottle crates, pipes and other extruded parts which have become unusable but which can be collected and recycled. Almost half a million tonnes of plastics materials are recycled each year from such sources as these. For these operations to be successful the materials have to possess an acceptably low level of contamination and the cost of collection and grading must not be so high that they make the process uneconomic. Such are the problems that make it difficult to recover plastics effectively from household waste.

In between incineration and recycling lies the possibility of taking plastics waste and chemically breaking it down into simpler chemicals by processes such as *pyrolysis*. This involves heating the plastics materials in the absence of air or oxygen, usually between 400° and 800°C. One commercial process yields valuable raw materials such as ethylene, propylene, methane, butadiene and benzene, many of which may themselves be used for plastics manufacture.

Benefits to the environment

The problems of litter, toxicity and fire are matters for concern, but the use of plastics has greatly benefited the standard and quality of life, and has also, on occasion, been a boon to the environment. Plastics piping, cabling, ducting and drainage may be hidden underground; the use of plastics thermal insulation leads to savings in heating fuels such as oil and coal and therefore results in less atmospheric pollution. A modern car may have many hundreds of plastics parts which have replaced heavier metal components, and the resulting savings in weight lead to reduced road wear, economies in fuel consumption and lower production of exhaust gases which pollute the atmosphere.

6 A Brief Survey of Plastics Materials

Well over a hundred types of plastics materials are commercially available. Although the market is dominated by four types which account for over three-quarters of the total tonnage produced – polyethylene, polypropylene, polystyrene and PVC – there are many others which can be found in most homes. These include the nylons, the polyacetals, ABS, urea-formaldehyde plastics and the polyurethanes. Before considering some of the more important types it is useful to summarise the main characteristics of plastics:

1 Most plastics may be processed by mass production techniques such as injection moulding which enables highly complex parts to be made in one operation with little need for assembly operations, and produces minimal scrap.

2 Colouring is usually distributed throughout the mass and not limited to the surface of the plastics.

3 Most plastics are excellent electrical and thermal insulators.

4 Plastics are available with a wide range of chemical and solvent resistances. Some dissolve in water, others are resistant to the strongest acids.

5 Plastics are much lighter in weight for a given volume (they have a lower density) than traditional materials such as glass and metals. Although not as strong as most metals, fibre-reinforced plastics often have a strength-per-unit weight as high as that of most metals.

6 Since most plastics may be fabricated at low temperatures, compared to those used for molten metals, the energy requirements for shaping are comparatively low.

Compared with most metals, plastics have certain limitations: they have more limited heat resistance and less stiffness and strength in particular. In many applications the limited heat resistance is not a problem and indeed the low softening point of many materials allows processing at low temperatures – most of the large tonnage thermoplastics are processed below 250°C. However, where heat resistance is required, special purpose materials are available which may be used at temperatures as high as 300°C.

Most plastics are less stiff and less strong than metals. Stiffness and strength can be increased considerably if they are reinforced with glass fibre or carbon fibre. Furthermore, because of the low density of such materials the strength-per-unit weight can be, in many cases, at least as high as that of metals.

In the rest of this section the most important plastics are briefly surveyed, beginning with polyethylene. This has the simplest structure of any polymer. It will then be shown how making small changes to this structure results in other important plastics such as polypropylene, polystyrene and polyvinyl chloride. Some other more specialised thermoplastics will then be reviewed, but due to their complexity their structure will not be dealt with here. The section concludes with a look at some important thermosetting plastics and some miscellaneous materials with rather special properties.

Polyethylene

In terms of tonnage consumption the polymer polyethylene (also known as polythene, polyethene and poly methylene) is the world's number one plastics material. In theory it also has the simplest structure consisting of a chain of carbon atoms with two hydrogen atoms attached to each carbon:

```
  H   H   H   H   H   H   H
  |   |   |   |   |   |   |
– C – C – C – C – C – C – C –
  |   |   |   |   |   |   |
  H   H   H   H   H   H   H
```

In practice, as explained above, the structure of commercial materials may be somewhat more complex.

Such a *hydrocarbon* structure has very good chemical resistance, is electrically a very good insulator and does not absorb water. The forces holding the molecules together are not very high, therefore polyethylene softens at quite low temperatures (85–125°C according to type) and does not have a very high tensile strength. The regular structure enables the polymer to crystallise partially, which explains why polyethylene is opaque in bulk. It is very tough, burns with a smoky flame and is one of a small number of solid plastics which float on water (polypropylene is another).

There are many grades of polyethylene on the market and these may be grouped into three main classes according to the shape of the polymer molecules. The original materials, first developed by ICI, consist of molecules comprising long chains to which are attached both long and short branches (figure 5a).

The short branches make it difficult for the molecules to pack together in a regular fashion and this makes crystallisation somewhat difficult. The inability to pack also means that there are fewer molecules in a given volume of space compared with more regular structures. This means that the density is reduced and such materials are usually known as low density polyethylene (LDPE). The LDPEs are generally the toughest, softest and the least heat-resistant of the various types of polyethylene. They are widely used to make packaging film, squeegee bottles, ink tubes for ball-point pens, and toys.

In the 1950s polyethylenes with unbranched molecules came onto the market (figure 5b). These crystallised to a greater extent, had a higher density and became generally known as high-density polyethylenes (HDPE) and occasionally as linear polyethylenes. Well-known uses include bottles for detergents, bleaches and white spirit, milk crates and fish boxes. More recently polyethylenes have become available with some short side chains but without long chains. These versions are known as linear low-density polyethylenes (LLDPE) (figure 5c) and are used in packaging film.

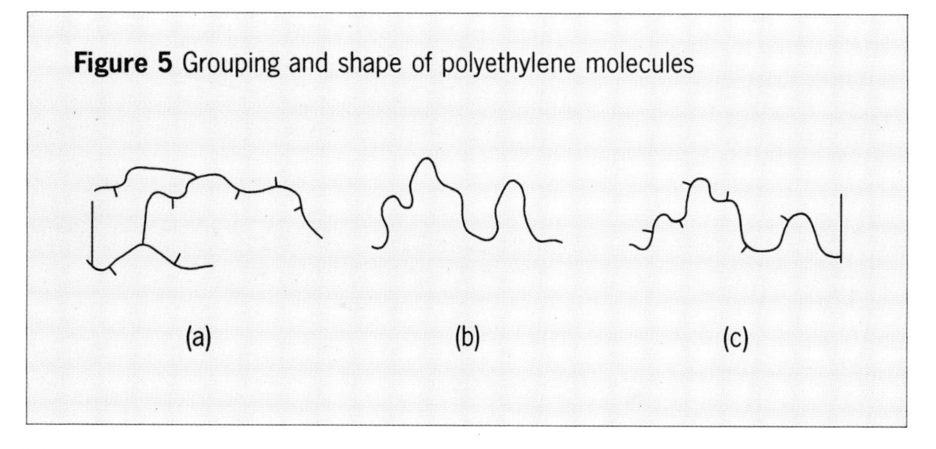

Figure 5 Grouping and shape of polyethylene molecules

Polypropylene

Polypropylene is closely related to polyethylene since it is one of a family of hydrocarbon polymers known as *polyolefins* or *polyalkenes*. It differs from polyethylene in that one of the hydrogen atoms on alternate carbon atoms in the chain has been replaced by a chemical group known as a methyl group, which consists of one carbon and three hydrogen atoms:

$$\begin{array}{cccccccccc} & H & H & H & H & H & H & H & \\ & | & | & | & | & | & | & | & \\ - & C & -C- & C & -C- & C & -C- & C & - \\ & | & | & | & | & | & | & | & \\ & CH_3 & H & CH_3 & H & CH_3 & H & CH_3 & \end{array}$$

The molecule is not symmetrical so that there are a number of possible structures according to the way in which the monomer molecules are linked together, and this determines the position of the methyl group. The materials normally encountered approximate to a regular form known as *isotactic polypropylene*.

Stereoregular polymers

In 1954 Giulio Natta worked in Milan on earlier work by the German chemist Karl Ziegler. Some interesting new catalyst systems were developed, for which Natta and Ziegler were awarded the Nobel Prize for chemistry. Not only were they able to polymerise monomers which had previously resisted conversion to polymers but in many cases it was found possible to produce a range of products from one monomer. This was particularly the case with monomers which differed from ethylene since one of the hydrogen atoms was replaced by another atom or group of atoms. Well-known examples of such monomers include propylene, styrene and vinyl chloride.

In order to understand these different structures it should be appreciated that atoms are not usually at an angle of 90° to each other as indicated in many of the diagrams in this booklet but more commonly at an angle slightly in excess of 100°. Thus the atoms cannot all lie in the plane of the paper and in polypropylene for example the chain could take the following forms where R represents the methyl (CH) group. The stretched zig-zag structures shown are themselves a simplification to aid comparison since in reality the polymers take the form of a spiral or helix.

In the diagrams the side groups which are joined by solid lines are in front of the plane of the paper whilst those joined by dotted lines are below the plane. The first two forms (isotactic and syndiotactic) have regular repeating units and the polymers can crystallise. The most important polypropylenes are the regular crystallisable isotactic polymers. In the third (atactic) form the arrangement is irregular and the polymer is amorphous. In the cases of PVC and polystyrene, where R represents either a chlorine atom or a benzene ring respectively, commercial materials are largely atactic.

Polyethylene and isotactic polypropylene have many similar properties since they are both crystallisable hydrocarbons. However polypropylene has a higher softening point (it will withstand boiling water for a limited period), is stiffer and mouldings are less wax-like in appearance. Some grades are rather brittle below 0°C, the freezing point of water.

The polymer is widely used for injection moulding articles such as stacking chairs, picnic ware, interior parts of cars, housewares, luggage and washing machine drums (when it is reinforced with glass fibre).

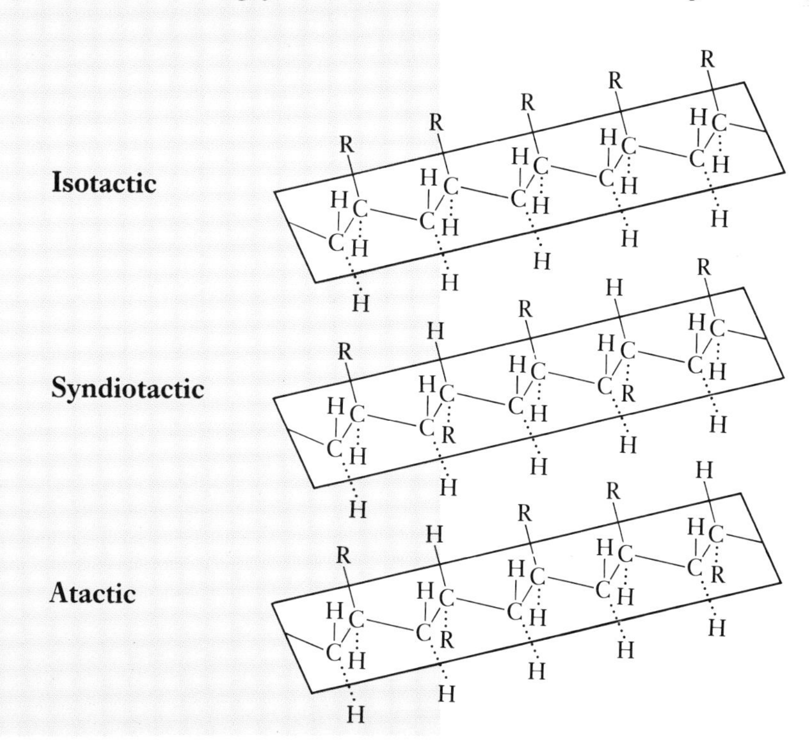

Polystyrene and high impact polystyrene

Polystyrene resembles polyethylene and polypropylene since it is also a hydrocarbon. It is also similar to polypropylene in that one of the hydrogen atoms on alternate carbon atoms in the main chain has been replaced by a larger chemical group, in this case a ring of six carbons and five hydrogens known as a benzene ring:

$$
\begin{array}{ccccccccccccccc}
 & H & & H & & H & & H & & H & & H & & H & \\
 & | & & | & & | & & | & & | & & | & & | & \\
- & C & - & C & - & C & - & C & - & C & - & C & - & C & - \\
 & | & & | & & | & & | & & | & & | & & | & \\
 & H & & \text{(O)} & & H & & \text{(O)} & & H & & \text{(O)} & & H &
\end{array}
$$

where (O) represents the benzene ring

Unlike polypropylene the normal commercial polystyrenes have an irregular arrangement of the side groups in relation to the main carbon chain and the material is not capable of crystallisation – it is said to be amorphous. Because there are no crystal structures to scatter light, unfilled polystyrene is transparent. (Expanded polystyrene, which is full of air, is opaque because light is scattered at the edge of the air bubbles in the polystyrene.) The presence of the benzene ring makes the chain stiff and polystyrene is a rigid material which gives off a metallic sound when struck. Whilst it has many uses as a rigid, transparent material (for example in packaging, toys and housewares) polystyrene also suffers from a number of disadvantages: it burns with a smoky flame; it is attacked, and in some cases dissolved by, many liquids; it does not withstand boiling water and it is somewhat brittle.

Brittleness may be reduced by blending the polystyrene with a synthetic rubber. The resulting material, known as high impact polystyrene (HIPS) or toughened polystyrene (TPS), is just as important as the unmodified polystyrene. The toughened material is used in refrigerator liners, toys and low cost appliance housings.

ABS

ABS is a complex terpolymer prepared from acrylonitrile, butadiene and styrene, the formulae for acrylonitrile and butadiene being:

$$
\begin{array}{l}
CH_2 = \underset{\displaystyle CN}{\underset{|}{CH}}
\end{array}
$$

Acrylonitrile

$$CH_2 = CH - CH = CH_2$$

Butadiene

where the N in the acrylonitrile represents a nitrogen atom.

ABS mouldings are rigid and tough. Moreover, if properly made they can have a finish superior to that of most other plastics and for this reason are widely used for housings of such domestic appliances as vacuum cleaners and food mixers, and in car parts such as radiator grilles, door

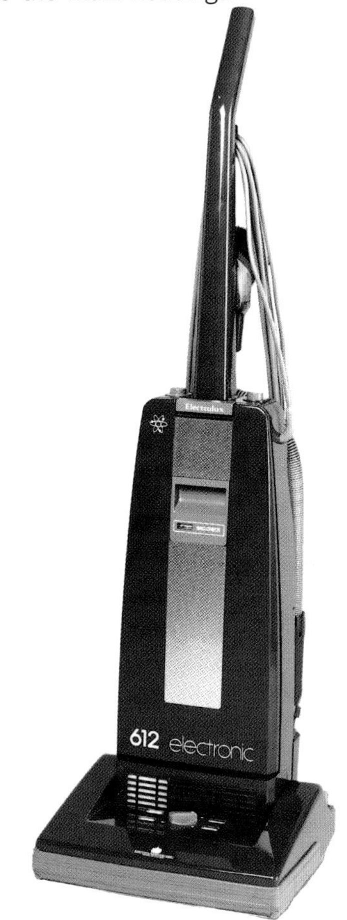

In vacuum cleaners plastics are used for the wheels, brush cylinders, suction fans and many other parts in addition to the main housings.

handles and fascia panels. Most grades are opaque and also have better heat resistance than polystyrene but they still do not withstand boiling water.

Polyvinyl chloride

Also known as PVC and polychlorethene, polyvinyl chloride is yet another example of a polymer in which one of the hydrogen atoms in the ethylene repeat unit is replaced by another group, in this case a chlorine atom:

$$
\begin{array}{ccccccccccccccc}
 & H & & H & & H & & H & & H & & H & & H & \\
 & | & & | & & | & & | & & | & & | & & | & \\
- & C & - & C & - & C & - & C & - & C & - & C & - & C & - \\
 & | & & | & & | & & | & & | & & | & & | & \\
 & H & & Cl & & H & & Cl & & H & & Cl & & H &
\end{array}
$$

(Cl represents a chlorine atom)

Early gramophone records were made of shellac. Modern records are usually made from vinyl chloride-vinyl acetate copolymers.

This makes the polymer somewhat harder than polyethylene and also more resistant to burning. It is also rather unstable on heating and for use has to be mixed with a stabiliser to prevent the PVC decomposing on heating during processing. There are two main types of PVC product, unplasticised PVC (UPVC) and plasticised PVC (PPVC).

UPVC is a blend of polymer, stabiliser, filler (such as china clay), pigment and processing aids. It is widely used in building for drain pipes and gutterings and in chemical plant. Thin film is used for magnetic tape and an unplasticised copolymer for gramophone records.

Plasticised PVC is used as the insulator for domestic electric flex, for leathercloth, for domestic hosepipes, playballs, baby pants, medical tubing and as cling-film wrapping material. In addition to the ingredients found in UPVC, plasticised PVC contains a compatible liquid known as a plasticiser. The softness of the material will depend on the amount of plasticiser present, and in some applications there may be as much plasticiser as polymer.

Many monomers contain the word vinyl in their name. Apart from vinyl chloride, other important materials include vinyl acetate (used to make polymers and copolymers for emulsion paints), vinyl fluoride and vinyl pyrrolidone. Styrene could be called vinyl benzene and propylene – vinyl methane. These monomers are so-called because they all contain the vinyl group $CH_2{=}CH{-}$.

Acrylic plastics
There are many types of acrylic polymers which find uses in areas such as paints, fibres and adhesives (including 'superglue') but when people refer to acrylic plastics they are usually talking about plastics from polymethyl methacrylate. This is best known in the sheet form *Perspex*. It has the formula:

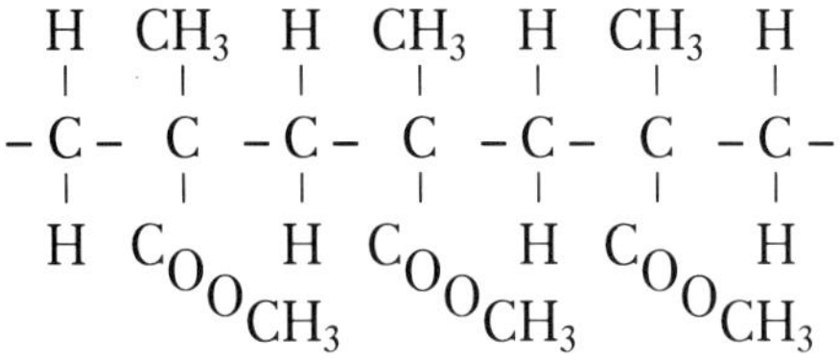

$$\begin{array}{ccccccc} \mathrm{H} & \mathrm{CH_3} & \mathrm{H} & \mathrm{CH_3} & \mathrm{H} & \mathrm{CH_3} & \mathrm{H} \\ | & | & | & | & | & | & | \\ -\mathrm{C}- & \mathrm{C} & -\mathrm{C}- & \mathrm{C} & -\mathrm{C}- & \mathrm{C} & -\mathrm{C}- \\ | & | & | & | & | & | & | \\ \mathrm{H} & \mathrm{COOCH_3} & \mathrm{H} & \mathrm{COOCH_3} & \mathrm{H} & \mathrm{COOCH_3} & \mathrm{H} \end{array}$$

where the O represents an oxygen atom.

This material has excellent resistance to weathering whilst unfilled grades are transparent, transmitting more light than most forms of glass. Common uses include display signs, road signs, 'plastic' baths and aircraft glazing. Injection mouldings are widely used for car rear-lamp housings whilst a more specialised compression moulding process is used to make false teeth. Copolymers closely related to polymethyl methacrylate are used to make soft contact lenses. The particular copolymer is chosen because of its ability to absorb water.

Nylons
(Also known as polyamides). Although better known as fibres and textiles the nylons are important engineering plastics, that is, they may be used in load-bearing applications. Not only are they tough and heat resistant but they resist hydrocarbon oils and abrasion better than many non-ferrous (iron-free) metals. Their biggest disadvantage is that they tend to absorb water, one important type by up to 14 per cent.

They are widely used for gear wheels, bearings, cams and bushes whilst monofilament is used for fishing lines, surgical sutures and brush tufting. By means of the process of polymerisation casting (in which polymerisation actually takes place in the mould) large gear wheels, conveyor buckets for the mining industries and propellers for small marine craft may be made.

Polyacetals
Also known as acetal resins, polyformaldehydes, polymethanals and polyoxymethylenes, these materials are similar to the nylons and are also widely used for gear wheels, bearings and related applications. Although they do not have quite the same abrasion resistance, they are more rigid and also have much lower levels of water absorption. They are also used in plumbing applications.

Polycarbonates
Polycarbonates have the useful property combination of toughness, transparency (with unfilled grades) and low water absorption (which also results in minimal staining by coloured liquids). I have had two drinking tumblers which have been used every day for nearly thirty years, most often with blackcurrant juice. These are still in good condition and quite unstained. Better known applications are to make the visors of astronauts' helmets and transparent riot shields. Other important uses include camera components, including camera bodies of many cameras; tough glazing, which is used for example in gymnasia or in vandal-sensitive areas; and appliance housings.

Polytetrafluoroethylene (PTFE)
In spite of its lengthy name this is structurally one of the simplest of plastics in which all of the hydrogen atoms in polyethylene have been replaced by fluorine atoms. The material is particularly well-known under the trade names *Fluon* and *Teflon*.

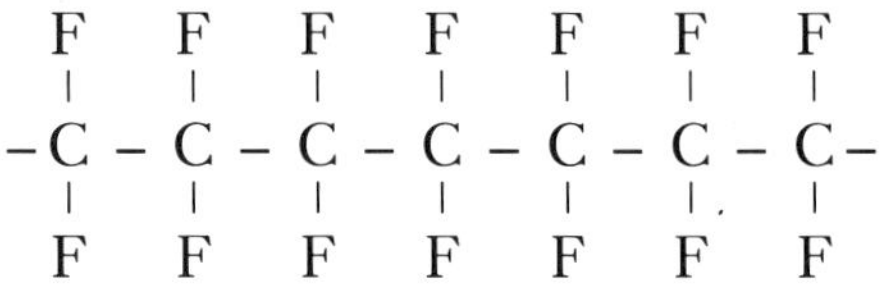

$$\begin{array}{ccccccc} \mathrm{F} & \mathrm{F} & \mathrm{F} & \mathrm{F} & \mathrm{F} & \mathrm{F} & \mathrm{F} \\ | & | & | & | & | & | & | \\ -\mathrm{C}- & \mathrm{C}- & \mathrm{C}- & \mathrm{C}- & \mathrm{C}- & \mathrm{C}- & \mathrm{C}- \\ | & | & | & | & | & | & | \\ \mathrm{F} & \mathrm{F} & \mathrm{F} & \mathrm{F} & \mathrm{F} & \mathrm{F} & \mathrm{F} \end{array}$$

PTFE is an exceptional material. It has a chemical resistance unmatched by any other plastics material and it is a superb electrical insulator. It shows very little friction when rubbed against other materials and is also renowned for its non-stick qualities. It also withstands temperatures above 250°C. It is more expensive than the other materials considered in this short review and it also

Squash courts made from acrylic sheet allow all-round viewing of a game. Earlier courts built with more traditional materials could not easily accommodate many spectators.

requires special expensive and time consuming methods of processing.

PTFE's combination of excellent properties results in it having many uses in chemical processing plants, in electrical equipment and in food processing. It is probably best known for its use in non-stick saucepans and other kitchen equipment.

Phenolic resins (phenol-formaldehyde plastics)

Unlike all of the materials described previously in this section the phenolics are thermosetting plastics, that is, they harden or cross-link on heating (unlike thermoplastics, which soften). Since chemical reactions take place during moulding, the moulding times take rather longer than for thermoplastics. Furthermore, scrap materials may not be re-used in the same way as they can with thermoplastics. Another disadvantage of phenolic resins is that they are usually dark in colour.

At one time the phenolics were widely used for electrical plugs and switches but the tendency for an electric current to burn a track along the surface of the moulding, particularly in damp conditions, has led to their replacement by other materials such as the UF plastics (see below) for this purpose. They are still of importance for making laminated plastics for electrical applications, and for the under-layers of decorative laminates. Thick fabric-reinforced laminates may be machined into a variety of non-metallic corrosion resistant products for engineering and other industries.

UF and other aminoplastics

The urea-formaldehyde (UF) plastics were introduced in the 1920s as thermosetting moulding materials available in a wide range of colours. Although less heat resistant than the phenolics they did not exhibit surface electrical tracking under damp conditions and most domestic plugs and switches in a typical home are made from this material.

At one time UFs were widely used for picnic-ware but they were not satisfactory because of their tendency to stain. Better results were obtained from melamine-formaldehyde (MF) plastics, a related aminoplastics material. Whilst largely replaced for this purpose by thermoplastics the MF resins are still very important for the manufacture of the surface layer of decorative laminates (although the cheaper phenolics are used for the underlayers).

Polyesters

There are a very large number of polymers that can be called polyesters. The name simply means that in the chain there are a number of links

Laminated plastics which are durable and easy to clean have revolutionised kitchen design. The normal decorative laminates are made from paper and phenolic resins with the upper layers using a decorative paper and melamine-formaldehyde resins which possess a superior abrasion resistance.

comprising a chemical group called an ester link that consists of one carbon and two oxygen atoms and may be written –COO–. The polyester *polyethylene terephthalate* is used to make the fibres Terylene and Dacron and the tough transparent bottles that are used for carbonated drinks such as lemonade, beers and colas. Most domestic gloss paints are polyester-based whilst glass-reinforced boats and car bodies are held together by polyester resin. Thermoplastic and thermosetting polyester moulding materials also exist.

Other plastics

The plastics briefly described above represent only a few, albeit the most important of plastics materials available. Where rigidity, toughness and heat resistance are required the designer might consider the polysulphones, aryl polymers, polyphenylene sulphides, polyamide-imides and so on. The heat-resistant thermoplastic polyether ether ketone (PEEK), reinforced with carbon fibre is used, for example, to produce helicopter tailplanes. For transparent uses styrene-acrylonitrile plastics, aromatic polyamides or even cellulose plastics might be considered. For water solubility there are polyvinyl alcohol, cellulose ethers and many other possible materials. Some materials are electrically activated by light (and find use in photocopying equipment). Others are piezoelectric, that is, they generate an electric current on application of pressure and are of interest in telephones and microphones. In microelectronics the ability to use plastics which harden on exposure to light is of considerable importance.

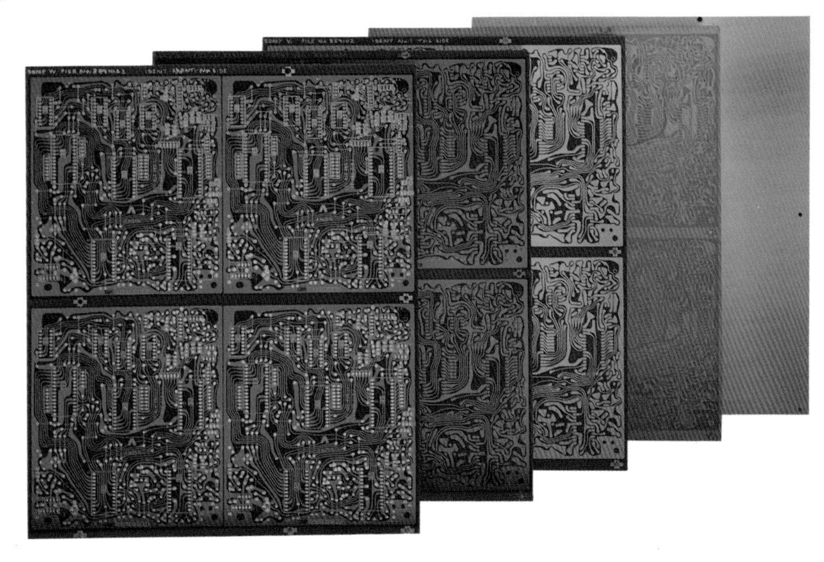

Printed circuit boards (PCBs) have been an important factor in the development of mass production methods in the electronics industries. A typical PCB is made by taking phenolic resin-paper laminate which has a copper layer on the surface; printing the circuit design on the copper and then etching away the copper not protected by the printing ink.

Sources of Further Information

The British Plastics Federation (BPF), 5 Belgrave Square, London SW1
The BPF is the organisation of the plastics industry. It represents the industry in discussion with government and other bodies and provides numerous services for its member companies.

Plastics Processing Industry Training Board (PPITB), Coppice House, Halesfield 7, Telford, Shropshire
The PPITB provides training support for the plastics industry. It offers training courses on such subjects as injection moulding, extrusion and blow moulding at its own Training Centre, advice on training to companies within the scope of the Board and also provides a number of training grants.

Plastics and Rubber Institute (PRI), 11 Hobart Place, London SW1W OHL
This organisation is the professional society for polymer scientists, technologists and engineers. It organises numerous conferences, seminars and other meetings, issues a number of publications including a bi-monthly journal and awards qualifications in polymer science and technology.

Science Museum, Exhibition Road, London SW7 2DD
The Science Museum has a representative collection of plastics dating from the 1850s to the present day.

Bibliography

Secondary school level (design and technology)
P.J. Clarke *Plastics for Schools*, Mills & Boon, 1976.
M.J.D. Hall *Design in Plastics*, Hodder & Stoughton, 1987.
D.H. Morton-Jones and J.W. Ellis *Polymer Products: Design Materials and Processing*, Chapman and Hall, 1986.
G.H. West *Manufacturing in Plastics*, Plastics and Rubber Institute (PRI), 1985.
G.H. West *Engineering Design in Plastics*, PRI, 1986.

Secondary school level (science)
D.A. Blackadder *Some Aspects of Basic Polymer Science*, Royal Society of Chemistry.
H.A. Finlay *Experiments in Polymer Chemistry*, Shell Educational Service (Shell-Mex House, Strand, London WC2R ODX).
P. Tooley *High Polymers*, John Murray, 1971.

Comprehensive reference books on plastics
R.D. Beck *Plastics Product Design*, Reinhold, New York, 1970.
J.A. Brydson *Plastics Materials* Butterworths, London, 5th Edition 1989.
J.M. DuBois *Plastics Mould Engineering*, Reinhold, New York, 1979.
P.C. Powell *Engineering with Polymers*, Chapman & Hall, London, 1983.

History of plastics
S. Katz *Early Plastics*, Shire Publications, 1986.
M. Kaufman *The First Century of Polymers*, PRI, London, 1963.

Glossary

This is a short glossary of terms that are frequently used in the text:

amorphous plastics Plastics materials in which the polymer used is non-crystalline, ie, has no definite form or shape. Amorphous plastics are usually transparent unless filled with additives or bubbles of air. Examples of amorphous plastics include PVC, polystyrene and acrylic plastics such as polymethyl methacrylate.

aromatic plastics Plastics which contain benzene rings in their structure, such as polystyrene and polycarbonates. Where the benzene ring forms part of the polymer chain backbone – as in polycarbonates, polyethylene terephthalate and the polysulphones – the chain is stiffened and the plastics soften at higher temperatures than where the ring is absent.

composites In the broadest sense this implies a product obtained by physically combining a polymer with something else such as another polymer, fillers, pigments or some sort of fibre-reinforcement. It is more commonly taken to mean a fibre-reinforced plastics material. Examples include paper-reinforced laminates such as Formica and Warerite, cotton fabric-reinforced laminates, glass fibre-reinforced plastics used to make such products as boats and sports cars, and carbon-fibre reinforced plastics used to make tennis racquets and of importance in aerospace applications.

copolymer These are polymers where two or more monomers are used to form the polymer chain. Various types of copolymer exist.

cross-linking This is the process of joining up polymer chains to form a three-dimensional network. The term is also used to cover the situation where a monomer contains more than two reactive groups and reacts to form a three-dimensional network without necessarily first forming a polymer chain.

crystalline plastics Plastics in which the polymer used is crystalline. In polymers crystallinity is usually much less perfect than in simple chemicals such as copper sulphate, common salt and Epsom salts. It is more accurate to picture zones in a plastic mass where the polymer molecules are arranged in a very orderly fashion intermixed with zones in which there is little order. Crystalline plastics are nearly always translucent or opaque. Examples include polyethylene, polypropylene, the nylons and the polyacetals.

homopolymer A polymer obtained using only one type of monomer (in contrast to a copolymer).

monomer A low molecular weight material which is polymerised to form a polymer. In the case of condensation and re-arrangement polymerisations (see section 3) these materials are often referred to as intermediates rather than monomers.

plastics It is almost impossible to define plastics accurately. The term usually refers to high molecular weight materials prepared synthetically, which at some stage are capable of flow in order that they can be shaped. There are naturally occurring plastics or those made by modifying natural materials (as with cellulose nitrate); and some, such as PTFE, flow only with difficulty. The above definition also includes material such as glasses, rubbers, fibres, adhesives and paints which are not normally considered as plastics.

polymer A chemical produced by joining together monomer molecules by the process of polymerisation.

polymerisation The process of joining together monomers or intermediates to produce polymers.

terpolymer This is an abbreviation for ternary copolymer and implies that three types of monomer have been used to make the polymer.

vinyl monomer A monomer containing a vinyl group $CH_2{=}CH–$.

Printed in the United Kingdom for HMSO.
Dd 291756 1/91 C30.